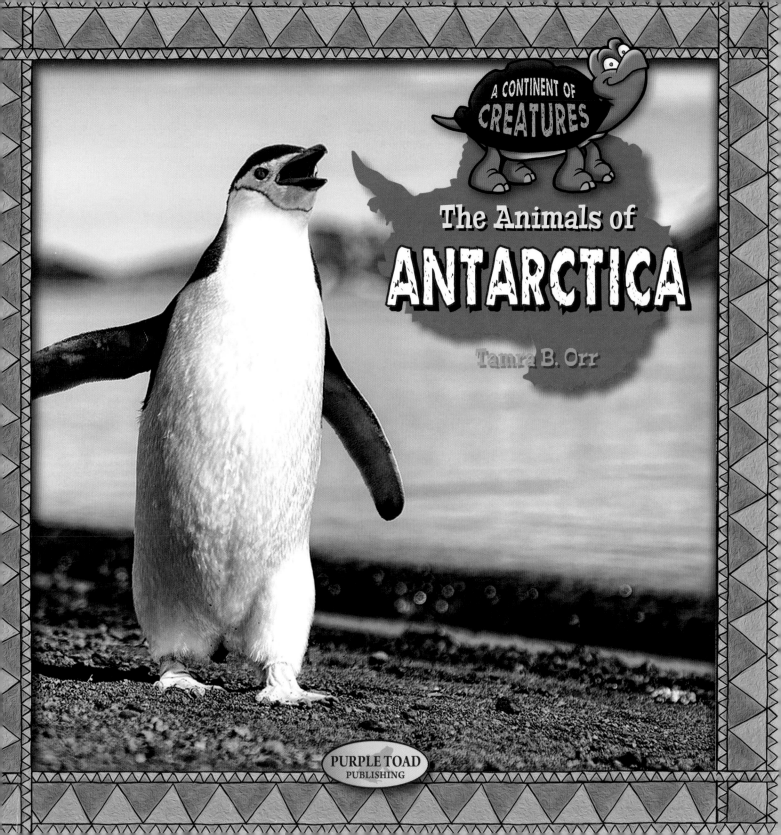

A CONTINENT OF CREATURES

The Animals of
ANTARCTICA

Tamra B. Orr

PURPLE TOAD
PUBLISHING

North America

Europe

Asia

Atlantic
Ocean

Africa

Indian
Ocean

Australia

Pacific
Ocean

South
America

Antarctica

Queen Maud
Land

ANTARCTICA

Ellsworth
Land

Wilkes
Land

Marie Byrd
Land

Victoria
Land

Welcome to Antarctica (ant-AR-tih-kah)! Brrrrr. Zip up your coat! This is the coldest, driest, and windiest biome in the world.

Antarctica surrounds the South Pole. It is the world's fifth largest continent.

Whales pop up to the top of the ocean to take a breath and greet any visitors.

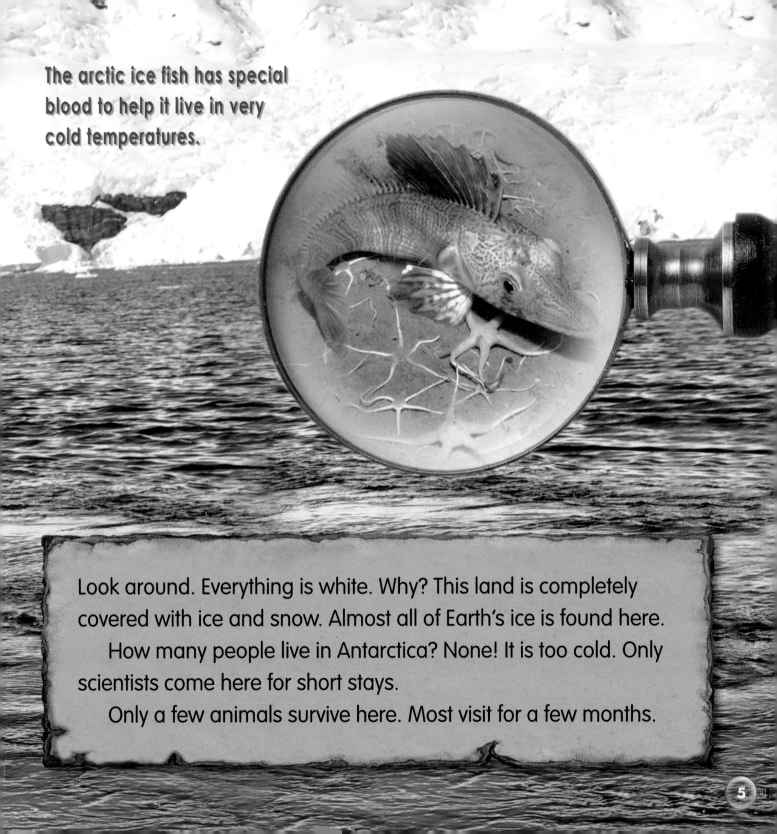

The arctic ice fish has special blood to help it live in very cold temperatures.

Look around. Everything is white. Why? This land is completely covered with ice and snow. Almost all of Earth's ice is found here.

How many people live in Antarctica? None! It is too cold. Only scientists come here for short stays.

Only a few animals survive here. Most visit for a few months.

The albatross's wingspan can be as much as 11 feet from tip to tip!

Look up! See those birds? Some birds come to Antarctica to make nests and lay eggs. The albatross can fly more than 500 miles in just one day. It uses the wind currents and almost never flaps its wings.

You can recognize a fulmar by its pink bill with a black tip.

The gray bird with black-tipped feathers is a fulmar. Don't go too close to it! These birds can spit a terrible smelling oil up to five feet. More than 100 million birds breed on the Antarctic coastline and its nearby islands.

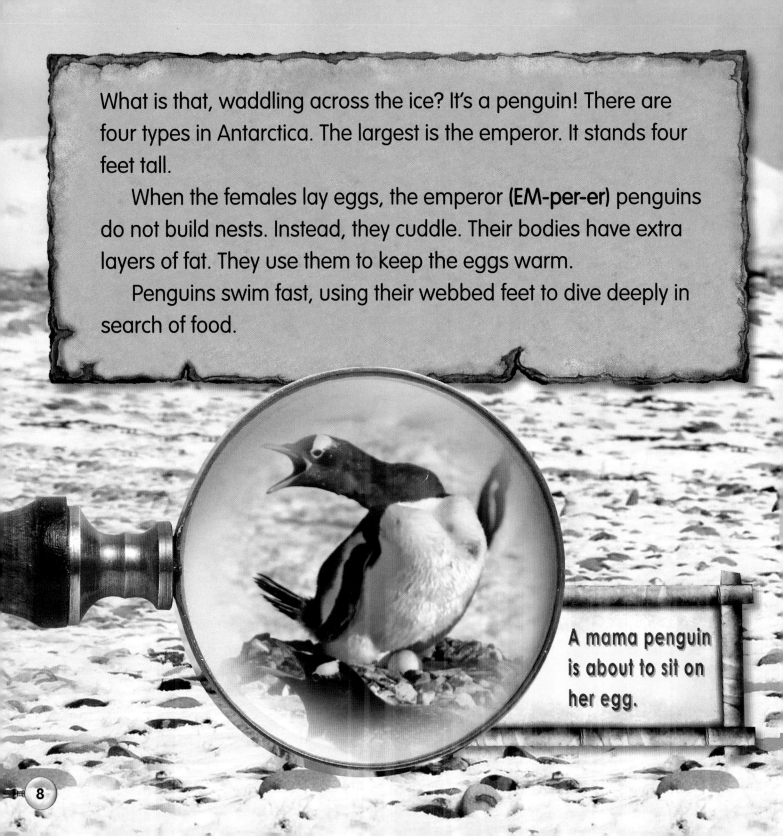

What is that, waddling across the ice? It's a penguin! There are four types in Antarctica. The largest is the emperor. It stands four feet tall.

When the females lay eggs, the emperor (EM-per-er) penguins do not build nests. Instead, they cuddle. Their bodies have extra layers of fat. They use them to keep the eggs warm.

Penguins swim fast, using their webbed feet to dive deeply in search of food.

A mama penguin is about to sit on her egg.

Emperor penguins have waterproof feathers and webbed feet.

Crabeater seals crawl out onto the ice every spring with their familes.

Six types of seals live in the Southern Ocean around Antarctica. Most are called crabeaters, even though they live on krill.

These seals are 10 feet long and weigh 500 pounds. In one year, they eat up to 25 times their body weight in tiny krill.

Krill are about
two inches long.

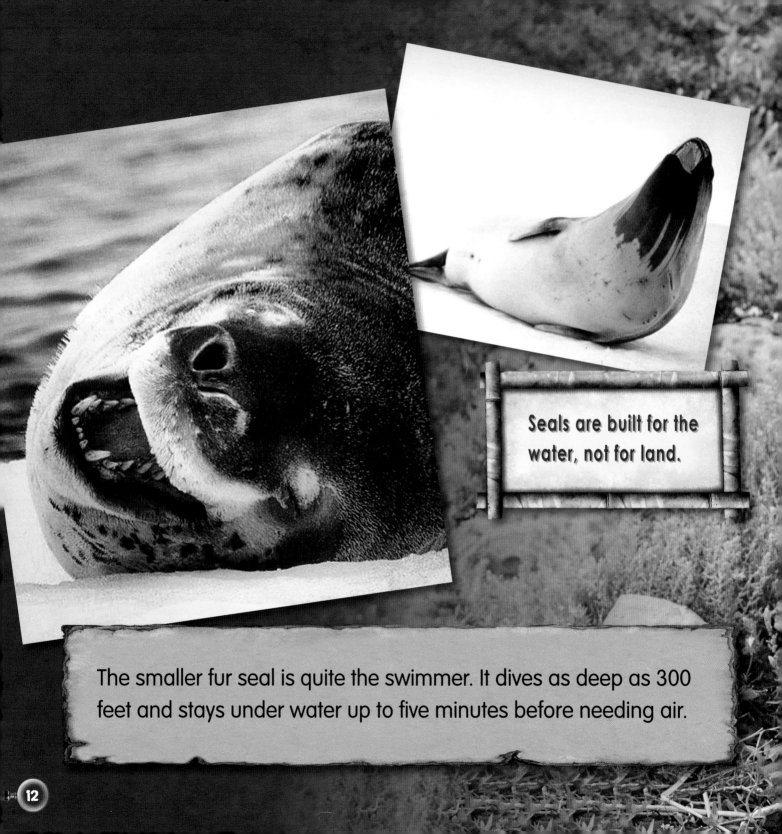

Seals are built for the water, not for land.

The smaller fur seal is quite the swimmer. It dives as deep as 300 feet and stays under water up to five minutes before needing air.

Baby sperm whales can be as long as 13 feet and weigh up to one ton.

The biggest animals in Antarctica are hidden beneath the waves. Many types of whales swim to the Southern Ocean.

Orca and sperm whales have teeth. They use them to grab fish, birds, and octopus. These types of whales are sometimes called "the wolves of the sea." They move in pods of up to 50 whales.

Mother and baby orca whales often stay together for at least a year.

Blue whales live in all the oceans of the world.

A scale model of a blue whale hangs in the American Museum of Natural History, in New York City.

The biggest animal on Earth spends time in Antarctic waters. The blue whale is 100 feet long (the length of a basketball court) and weighs 200 tons. When it comes to the surface of the ocean and exhales, the spray from its blowhole soars as high as 30 feet!

This tiny midge fly is considered the largest land animal in Antarctica.

Sea spiders are the size of a dinner plate. They are part of the crab family.

One of the only insects here is the Antarctic midge or fly. It spends most of its life frozen in ice. It is less than a quarter inch long and lives for only about a week.

There are spiders in the Antarctic, but you won't find them on the ice. They live in the sea looking for food.

The king crab lives on the ocean floor eating smaller creatures.

The "stars" of the sea are like this pink sea star. It will eat almost anything it can find in the ocean.

The cold water has kept the king crab from the Southern Ocean for millions of years. Warmer waters are now allowing them to live there, and they have become a threat to many creatures.

Mom and fur seal pup are clearly happy in this cold world!

The ice and snow of Antarctica are not comfortable for people. But for certain animals, it truly is Home, Sweet Frozen Home.

When early explorers first saw penguins, they thought they were fish. They are birds but can't fly.

FURTHER READING

Works Consulted

Cowcher, Helen. *Antarctica*. New York: Square Fish, 2009.

Jenkins, Martin. *The Emperor's Egg*. Paradise, CA: Paw Prints Press, 2009.

Kurkov, Lisa. *Icy! Antarctica*. Nigeria: Spectrum, 2014.

Saxton, Liam. *A Smart Kids Guide to Abundant Antarctica*. Thought Junction Publishing, 2015.

Viva, Frank. *A Trip to the Bottom of the World with Mouse*. Jackson, TN: Toon Books, 2012.

Web Sites

Antarctica at Time for Kids offers readers almost two dozen articles, videos, interactive activities and stories about this frozen land.

http://www.timeforkids.com/minisite/antarctica

Australian government's site: you can see live pictures from any of Australia's six webcams positioned in the Antarctic.

http://www.antarctica.gov.au/webcams

Crittercam: Antarctic Adventure gives kids the chance to take an interactive ride through Antarctica in this game from National Geographic.

http://kids.nationalgeographic.com/kids/games/geographygames/crittercamantarctic/

albatross (AL-bah-tross)—A large Antarctic bird.

biome (BY-ohm)—A community of animals and plants living together in a specific climate.

blowhole (BLOH-hole)—A hole on top of a whale's head used for breathing.

breed (breed)—To mate and produce young.

continent (KON-teh-nent)—One of the seven large land masses of the Earth.

cormorant (KOR-mor-ant)—A dark-colored bird with a long neck.

current (KER-ent)—A constant movement of water in the same direction.

exhale (EX-hayl)—To breathe out.

krill (KRIL)—Very small creatures in the ocean.

larva (LAR-vah)—A very young form of an insect.

petrel (PEH-tril)—A bird with long wings and dark feathers.

pod (POD)—a group of ocean animals.

INDEX

Printing   1   2   3   4   5   6   7   8   9

The Animals of Africa
The Animals of Antarctica
The Animals of Asia
The Animals of Australia
The Animals of Europe
The Animals of North America
The Animals of South America

ABOUT THE AUTHOR: Tamra Orr is the author of hundreds of books for readers of all ages. She loves the chance to learn about faraway lands and see what it is like to live there—all from the comfort of her work desk. Orr is a graduate of Ball State University, and is the mother of four. She lives in the Pacific Northwest and goes camping whenever she gets the chance.

Publisher's Cataloging-in-Publication Data
Orr, Tamra.
  Antarctica / written by Tamra Orr.
    p. cm.
Includes bibliographic references, glossary, and index.
ISBN 9781624692765
1. Animals—Antarctica—Juvenile literature. I. Series: A continent of creatures.
  QL106 2017
  591.998

eBook ISBN: 9781624692772

Library of Congress Control Number: 2016937182